When Science Meets the Quran

Messaoudi Abou nouh

Published by Messaoudi Abou nouh, 2026.

WHEN SCIENCE MEETS THE QURAN

First edition. February 9, 2026.

ISBN: 979-8233843808

Written by Messaoudi Abou nouh.

Also by Messaoudi Abou nouh

UIlu _al_azm
The Story of Noah (Peace Be Upon Him)
Prophet Abraham story

Standalone
Prophet Yusuf (Josef): A Story of Patience, Faith, and Divine Wisdom
When Science Meets the Quran

To my beloved parents,

whose prayers, patience, and sacrifices shaped my path.

To my wife,

my constant support, my strength in difficult moments,

and my partner in every step of this journey.

To my children,

my greatest inspiration and my deepest hope for the future.

May this work remind you that knowledge, faith, and perseverance

can illuminate even the longest roads.

— Messaoudi Abou Nouh

When Science Meets the Quran

Modern Discoveries Reflected in Ancient Revelation

Messaoudi Abou nouh

Published independently.

First Edition – 2026

Dedication

To my beloved parents,
whose prayers, patience, and sacrifices shaped my path.
To my wife,
my constant support, my strength in difficult moments,
and my partner in every step of this journey.
To my children,
my greatest inspiration and my deepest hope for the future.
May this work remind you that knowledge, faith, and perseverance
can illuminate even the longest roads.
— Messaoudi Abou Nouh

Introduction

For centuries, humanity has sought truth—through philosophy, science, and faith. Each generation has sought answers to the same fundamental questions:

Where did we come from?

How does the universe truly work?

Is there a deeper order behind what we observe?

Science, through observation and experimentation, has gradually uncovered many of the laws governing the universe. Yet, with every discovery, new questions emerge, reminding us that knowledge is a journey, not a destination.

The Quran, revealed over fourteen centuries ago, is primarily a book of guidance—calling humanity toward reflection, morality, and awareness of the Creator. Remarkably, within its verses are references to natural phenomena that were far beyond the scientific understanding of the time in which it was revealed. These references were not presented as scientific textbooks, but as signs—invitations to think, to observe, and to reflect.

This book does not seek to force conclusions, nor does it aim to replace scientific research or academic inquiry. Instead, its purpose is to explore points of convergence between modern scientific discoveries and verses found in the Quran—connections that invite honest reflection and thoughtful discussion.

The goal of this work is simple yet profound:

to search for truth with sincerity,

to present knowledge in a clear and Chapter Three

Mountains: Deep Roots Beneath the Earth

For most of human history, mountains were seen simply as massive formations rising above the Earth's surface—symbols of strength, beauty, and permanence. Their true structure, however, remained hidden. Only in recent centuries did science uncover a remarkable fact: mountains extend far deeper beneath the surface than what the eye can see.

A Quranic Description Ahead of Its Time

The Quran makes a striking reference to mountains in Surah An-Naba (78:6–7):

"Have we not made the earth a resting place,

and the mountains as stakes?"

The word translated as stakes (awtād) refers to objects driven into the ground to secure and stabilize—such as tent pegs or anchors. This description implies depth, firmness, and a stabilizing function.

At the time of the Quran's revelation, there was no scientific framework capable of understanding the internal structure of mountains. Yet the metaphor used is remarkably consistent with what modern geology has since revealed.

Modern Geological Discovery

With the development of geology and geophysics, scientists discovered that mountains possess deep subterranean roots extending into the Earth's crust—sometimes several times deeper than their visible height.

This phenomenon is explained through the principle of isostasy, which describes how the Earth's crust floats in balance atop the denser mantle beneath it. Just as an iceberg has a large hidden mass below water to support what appears above, mountains rely on deep roots to maintain structural equilibrium.

Key scientific findings include:

Deep Roots: Mountain ranges such as the Himalayas have roots that extend tens of kilometers into the crust.

Crustal Balance: These roots help distribute pressure and maintain balance within the Earth's lithosphere.

Tectonic Formation: Mountains often form at the boundaries of tectonic plates, where immense forces push the crust upward while simultaneously thickening it below the surface.

These discoveries emerged primarily in the 19th and 20th centuries—long after the Quran had described mountains using a term that precisely conveys this hidden anchoring structure.

Precision Without Excess

It is important to approach this topic with accuracy. Mountains do not prevent earthquakes entirely, nor do they stabilize the entire planet in an absolute sense. However, they do play a significant structural role within the Earth's crust, contributing to local and regional balance.

The Quran's use of the term stakes does not overstate the function of mountains—it describes it with functional precision, leaving room for scientific understanding to unfold over time.

An Invitation to Reflect

What makes this correspondence remarkable is not only the accuracy of the description, but its restraint. The Quran does not explain geological processes in technical language; instead, it offers imagery that conveys meaning without scientific overreach.

For the reader, this convergence raises a thoughtful question:

How could a text revealed more than fourteen centuries ago employ terminology that aligns so closely with geological concepts discovered only in recent history?

Conclusion

Mountains are far more than surface features. They are complex structures with deep foundations, silently shaping the stability of the land we inhabit. Modern science uncovered this reality through centuries of observation and measurement.

The Quran, however, referred to this hidden truth with a single, powerful word—stakes—inviting generations to observe, reflect, and seek understanding.

In this harmony between ancient revelation and modern discovery, we are reminded that some truths remain concealed until humanity develops the tools to recognize them.accessible way, and to guide readers—especially those curious about the Quran—toward understanding its message through a rational and scientific lens.

Each chapter examines a specific scientific discovery that has been established through modern research, then reflects on how similar concepts are referenced in the Quran. The approach is analytical, respectful, and grounded in verified knowledge, avoiding speculation or exaggeration.

Whether you are a believer seeking deeper understanding, a student of science, or a reader simply searching for truth beyond assumptions, this book invites you to read with an open mind. The intention is not to convince through emotion, but to illuminate through reason.

Truth does not fear examination.

And sincere inquiry is the first step toward understanding.

Chapter One

The Creation of Human Life: Embryology in the Quran

For centuries, the origin and development of human life remained one of the greatest mysteries in science. Without advanced instruments, microscopes, or medical imaging, early civilizations could only speculate about what occurred inside the womb. Accurate knowledge of embryonic development was simply beyond human reach.

Yet, within the Quran—revealed more than fourteen centuries ago—some verses describe the stages of human development with remarkable clarity and precision.

Quranic Description of Human Development

In Surah Al-Mu'minun (23:12–14), the Quran states:

"We created man from an extract of clay.

Then we placed him as a drop (nutfah) in a secure place.

Then We made the drop into a clinging form (alaqah),

And we made the clinging form into a lump (mudghah),

And we made the lump into bones,

and we clothed the bones with flesh,

then We produced him as another creation.

So blessed is God, the Best of creators."

These verses outline a clear sequence of developmental stages, each described using precise terms that reflect observable biological realities.

Modern Scientific Understanding

Modern embryology confirms that human development occurs in distinct stages:

Nutfah (Drop):

The fertilized ovum, formed when sperm and egg unite, marks the beginning of human life.

Alaqah (Clinging Form):

At this stage, the embryo attaches itself to the uterine wall. Remarkably, the term alaqah conveys both the idea of clinging and a leech-like appearance—an accurate description of the embryo's form and behavior during early implantation.

Mudghah (Chewed-like Lump):

The developing embryo begins to resemble a small mass with somite segments, similar in appearance to a chewed substance—precisely reflected by the Quranic term mudghah.

Formation of Bones and Flesh:

Scientific research shows that cartilage structures appear first, followed by muscle tissue forming around them—aligning closely with the Quranic description of bones being formed and then clothed with flesh.

Another Creation:

This stage corresponds to the emergence of distinct human features and the complex development of organs and systems, marking a transition from basic biological structure to a fully recognizable human form.

A Remarkable Precision

What makes these descriptions extraordinary is not only their accuracy, but their timing. At the time of the Quran's revelation, there were no scientific tools capable of observing embryonic development in such detail. Even ancient medical theories contained significant inaccuracies when compared to modern knowledge.

The Quran does not present these stages as scientific claims, but as signs—inviting reflection on the precision and order embedded within human creation.

Between Faith and Science

This convergence between Quranic verses and modern embryology does not seek to replace science, nor does it demand blind belief. Instead, it opens a space for thoughtful inquiry.

For readers who value evidence, these verses raise an important question:

How could such accurate descriptions of embryological stages appear in a text revealed over fifteen centuries ago?

The answer to this question is left to the reader's reflection. What is undeniable, however, is that the Quran's description of human development aligns with scientific discoveries that humanity only confirmed in recent history.

Reflection

Human life begins unseen, fragile, and dependent—yet governed by precise laws and stages. Whether viewed through the lens of science or faith, the journey from a single cell to a complete human being remains one of the most profound processes in existence.

In this process, the Quran offers not only insight, but an invitation:

to observe, to think, and to seek truth without fear.

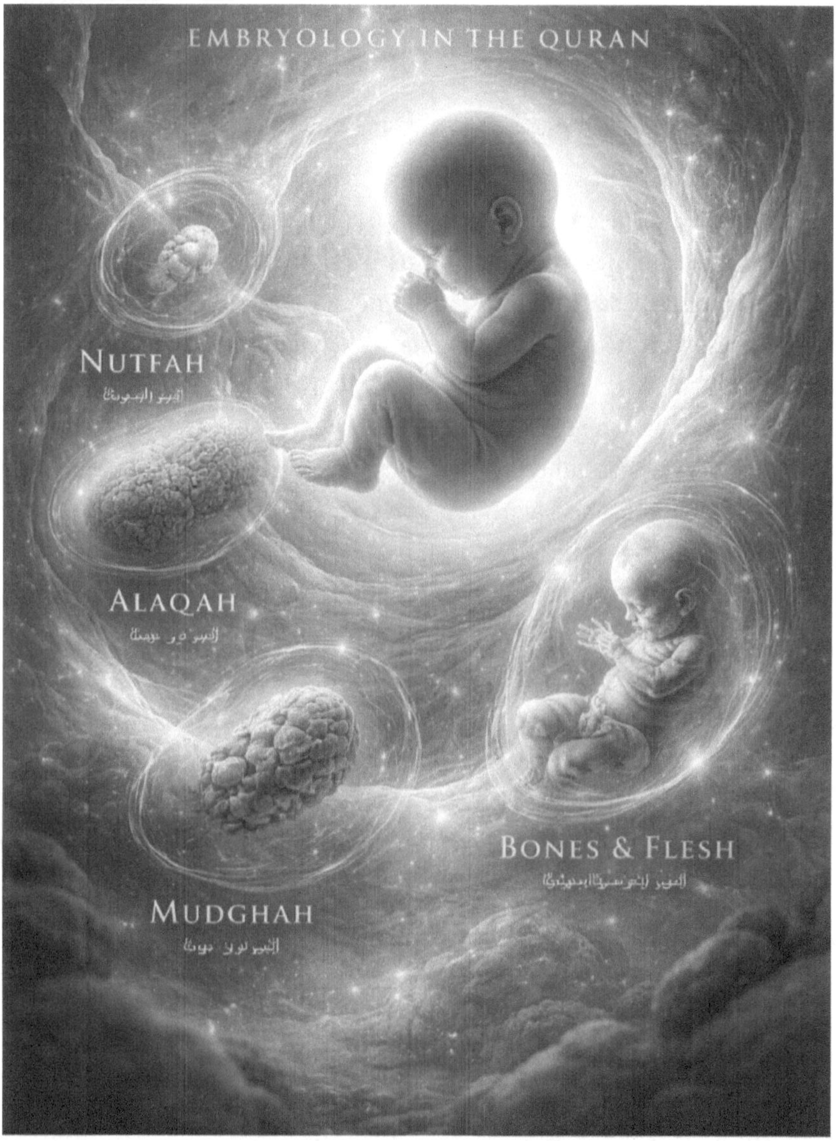
EMBRYOLOGY IN THE QURAN
NUTFAH
ALAQAH
MUDGHAH
BONES & FLESH

Chapter Two

The Universe and the Origin of Creation

The question of how the universe began has fascinated humanity for millennia. Ancient civilizations speculated about the heavens, the stars, and the vastness of the cosmos, but lacked the means to observe it systematically. Today, modern astronomy and cosmology have revealed astonishing truths about the origins and structure of the universe, yet questions remain that challenge even the greatest minds.

Remarkably, the Quran, revealed over fourteen centuries ago, contains verses that allude to these cosmic realities, describing creation in ways that align closely with modern scientific understanding.

Quranic Insights into the Cosmos

Several verses in the Quran speak about the origin and expansion of the universe:

Surah Al-Anbiya (21:30):

"Do not those who disbelieve see that the heavens and the earth were a joined entity, and We separated them?"

Surah Adh-Dhariyat (51:47):

"And the heaven We constructed with might, and indeed We are [its] expander."

These verses suggest a primordial unity of the heavens and the earth and a continuous expansion of the universe—concepts that modern cosmology has only confirmed in the last century.

Modern Scientific Perspective

The Big Bang:

Modern science explains that the universe originated from an extremely hot and dense point roughly 13.8 billion years ago. This initial state, expanding over time, led to the formation of galaxies, stars, and planets. The Quranic phrase "joined entity" aligns strikingly with this understanding.

Cosmic Expansion:

Observations by Edwin Hubble and subsequent research show that the universe is continuously expanding. The Quran's description of the heavens being "expanded" is a remarkable anticipation of this modern discovery.

Order and Precision:

The arrangement of celestial bodies—the rotation of planets, the balance of gravitational forces, and the motion of stars—reflects a system of precision and order. The Quran frequently reminds humanity to observe the heavens and reflect, which resonates with the modern approach of scientific inquiry.

A Call to Reflection

The Quran does not present itself as a textbook of astronomy. Instead, its verses invite observation, reflection, and thoughtful inquiry. The precision and timing of these statements—made over a millennium before the tools of modern astronomy—illustrate a profound awareness of natural phenomena that were otherwise unknowable at the time.

By examining the universe through both scripture and science, readers can appreciate the depth of knowledge encoded in the Quran and explore the harmony between faith and reason.

Conclusion

The universe, in its vastness and complexity, continues to inspire awe. From the cosmic explosion that marked its beginning to the intricate balance of galaxies and stars, the Quran provides a lens through which we can contemplate creation with curiosity and humility.

Through this chapter, we see that human understanding of the cosmos is a journey—a journey that was subtly guided in ancient revelation long before telescopes and satellites made the unseen visible.

THE UNIVERSE
AND THE ORIGIN OF CREATION
Surah Al-Anbiya (21:30)
"Do not those who disbelenve see that the heavens and the earth were a joined entity, and We separated them?
Surah Adh-Dhariyat (51:47)
"And the heaven We constructed with might, and inde ed We are [its] expander."
THE QURAN & MODERN SCIENCE

Chapter Three

Mountains: Deep Roots Beneath the Earth

For most of human history, mountains were seen simply as massive formations rising above the Earth's surface—symbols of strength, beauty, and permanence. Their true structure, however, remained hidden. Only in recent centuries did science uncover a remarkable fact: mountains extend far deeper beneath the surface than what the eye can see.

A Quranic Description Ahead of Its Time

The Quran makes a striking reference to mountains in Surah An-Naba (78:6–7):

"Have we not made the earth a resting place,

and the mountains as stakes?"

The word translated as stakes (awtād) refers to objects driven into the ground to secure and stabilize—such as tent pegs or anchors. This description implies depth, firmness, and a stabilizing function.

At the time of the Quran's revelation, there was no scientific framework capable of understanding the internal structure of mountains. Yet the metaphor used is remarkably consistent with what modern geology has since revealed.

Modern Geological Discovery

With the development of geology and geophysics, scientists discovered that mountains possess deep subterranean roots extending into the Earth's crust—sometimes several times deeper than their visible height.

This phenomenon is explained through the principle of isostasy, which describes how the Earth's crust floats in balance atop the denser mantle beneath it. Just as an iceberg has a large hidden mass below water to support what appears above, mountains rely on deep roots to maintain structural equilibrium.

Key scientific findings include:

Deep Roots: Mountain ranges such as the Himalayas have roots that extend tens of kilometers into the crust.

Crustal Balance: These roots help distribute pressure and maintain balance within the Earth's lithosphere.

Tectonic Formation: Mountains often form at the boundaries of tectonic plates, where immense forces push the crust upward while simultaneously thickening it below the surface.

These discoveries emerged primarily in the 19th and 20th centuries—long after the Quran had described mountains using a term that precisely conveys this hidden anchoring structure.

Precision Without Excess

It is important to approach this topic with accuracy. Mountains do not prevent earthquakes entirely, nor do they stabilize the entire planet in an absolute sense. However, they do play a significant structural role within the Earth's crust, contributing to local and regional balance.

The Quran's use of the term stakes does not overstate the function of mountains—it describes it with functional precision, leaving room for scientific understanding to unfold over time.

An Invitation to Reflect

What makes this correspondence remarkable is not only the accuracy of the description, but its restraint. The Quran does not explain geological processes in technical language; instead, it offers imagery that conveys meaning without scientific overreach.

For the reader, this convergence raises a thoughtful question:

How could a text revealed more than fourteen centuries ago employ terminology that aligns so closely with geological concepts discovered only in recent history?

Conclusion

Mountains are far more than surface features. They are complex structures with deep foundations, silently shaping the stability of the land we inhabit. Modern science uncovered this reality through centuries of observation and measurement.

The Quran, however, referred to this hidden truth with a single, powerful word—stakes—inviting generations to observe, reflect, and seek understanding.

In this harmony between ancient revelation and modern discovery, we are reminded that some truths remain concealed until humanity develops the tools to recognize them.

MOUNTAINS
AND DEEP ROOTS BENEATH THE EARTH
Surah An-Naba (78:7)
And the mountains as stakes?
THE QURAN & MODERN SCIENCE

Chapter Four: The Expansion of the Universe

For centuries, humanity believed that the universe was static and unchanging — a vast, fixed space in which stars and galaxies remained in their places. This view dominated scientific thought until the early twentieth century, when a revolutionary discovery changed our understanding of the cosmos forever.

A Universe in Motion

In 1929, astronomer Edwin Hubble made a groundbreaking observation: distant galaxies were moving away from us. Even more astonishing, the farther a galaxy was, the faster it appeared to be receding. This discovery led scientists to a profound conclusion — the universe is expanding.

This expansion does not mean that galaxies are simply moving through empty space like objects drifting apart. Instead, space itself is expanding, causing the distances between galaxies to increase over time. The universe is not static; it is dynamic and continuously growing.

How Do Scientists Know?

The key evidence for the expanding universe comes from a phenomenon known as redshift. When light travels from distant galaxies toward Earth, its wavelength stretches, shifting toward the red end of the spectrum. This stretching occurs because the space through which the light travels is itself expanding.

Redshift measurements consistently show that nearly all galaxies are moving away from one another, confirming that cosmic expansion is happening in every direction — with no central point.

No Center, No Edge

One of the most remarkable aspects of this discovery is that the universe has no center. Every observer, regardless of location, would see other galaxies moving away. This means the expansion is universal, not directional.

A helpful analogy is a balloon being inflated with dots drawn on its surface. As the balloon expands, all dots move farther apart, yet none of them is the center of the expansion. The surface itself stretches — just as space does in our universe.

A Qur'anic Perspective

More than fourteen centuries ago, the Qur'an made a statement that resonates strikingly with modern cosmology:

"And the heaven We constructed with strength, and indeed, We are expanding it."

(Qur'an, 51:47)

The Arabic expression used in this verse indicates ongoing action, suggesting a continuous process rather than a completed event. At a time when humanity believed the heavens to be fixed and eternal, the Qur'an described the universe as something that is actively expanding.

A Remarkable Convergence

It is important to emphasize that the Qur'an is not a science textbook, nor does it aim to replace scientific inquiry. However, the alignment between this ancient statement and a discovery made only in the last century invites reflection.

The expansion of the universe is now a foundational principle of modern cosmology, supported by extensive observational evidence. That such a concept appears in a text revealed long before the invention of telescopes or modern physics raises profound questions for those who seek truth with an open mind.

An Invitation to Reflect

This convergence between revelation and discovery does not demand blind belief. Instead, it offers an invitation — to think, to question, and to explore the origins of knowledge. In the vastness of an ever-expanding universe, humanity continues its search for meaning, guided by both reason and reflection.

THE EXPANSION
OF THE UNIVERSE
“And the heaven We constructed with strength,
and indeed, We are expanding it.”
Qur'an 51:47

Chapter Five: The Barrier Between Seas

For a long time, it was assumed that when two bodies of water meet, they mix completely and become a single body of water. This assumption seemed logical and uncontested for centuries. However, modern oceanography has revealed a surprising reality — not all seas fully merge, even when they meet side by side.

Meeting Without Mixing

In several locations around the world, scientists have observed that bodies of saltwater and freshwater, or seas with different salinity levels, can meet while maintaining distinct physical properties. These waters flow alongside one another, yet each preserves its own composition.

This phenomenon is not visible to the naked eye, but it is scientifically measurable and well-documented. The separation occurs due to differences in salinity, temperature, density, and pressure, creating what scientists refer to as a natural boundary zone.

The Scientific Explanation

Oceanographers refer to this boundary as a halocline — a layer where water salinity changes sharply. In some cases, this boundary acts as a barrier that limits mixing between two bodies of water.

Even though waves and currents exist, complete blending does not occur immediately. Instead, the waters remain distinct for long distances and extended periods of time. Advanced instruments, satellite imagery, and underwater exploration have confirmed this phenomenon beyond doubt.

A Statement from the Qur'an

Remarkably, the Qur'an describes this reality with striking clarity:

"He released the two seas, meeting side by side; between them is a barrier they do not cross."

(Qur'an 55:19–20)

Another verse states:

"And it is He who has let free the two seas, one fresh and sweet, and the other salty and bitter, and He placed between them a barrier and a forbidden partition."

(Qur'an 25:53)

These verses describe not only the meeting of different waters but also the presence of an invisible barrier that prevents them from overwhelming one another.

Knowledge Beyond Its Time

At the time of the Qur'an's revelation, humanity possessed no scientific tools to observe underwater boundaries, measure salinity gradients, or analyze oceanic layers. The seas appeared uniform and indistinguishable beneath the surface.

The concept of a stable boundary between waters of different characteristics would have been impossible to verify — or even imagine — without modern scientific methods.

Reflection and Meaning

Once again, the Qur'an does not present itself as a textbook of science. Yet its descriptions align with discoveries made many centuries later. This alignment invites reflection, not coercion — encouraging the reader to consider the source of such knowledge.

The invisible barriers within the seas stand as a reminder that nature operates according to precise laws, many of which humanity continues to uncover only gradually.

An Open Invitation

For the sincere seeker of truth, such correspondences raise thoughtful questions. How could a text revealed in the seventh century describe natural phenomena that remained hidden until the modern age?

The answer is left to the reader — guided by reason, observation, and an open mind.

THE BARRIER
BETWEEN SEAS
Qur'an 55:19:20 - Between them is a barrier they do not cross.

Chapter Six

Life Originates from Water

For much of human history, the origin of life remained one of the greatest unanswered questions. Philosophers, scientists, and scholars offered theories, yet none could fully explain how life began or what conditions made it possible. Only in recent centuries has modern science reached a clear and undeniable conclusion: water is essential for life.

Water as the Foundation of Life

Every known form of life on Earth depends on water. From the smallest microorganism to the largest living organism, water plays a central role in biological structure and function.

Modern biology confirms that:

Water acts as a universal solvent, allowing vital chemical reactions to occur.

Cells—the basic units of life—are composed largely of water.

Essential biological processes such as metabolism, nutrient transport, and temperature regulation require water to function.

Without water, life as we know it cannot exist.

A Qur'anic Statement of Profound Simplicity

More than fourteen centuries ago, the Qur'an made a statement that resonates powerfully with modern scientific understanding:

"And We made from water every living thing."

(Qur'an 21:30)

This verse does not present a complex explanation, nor does it list scientific details. Instead, it conveys a foundational truth with striking clarity: life is inseparable from water.

At the time of the Qur'an's revelation, such a statement could not have been supported by experimental science. The microscopic world of cells, biochemical reactions, and molecular biology was entirely unknown.

Modern Scientific Confirmation

Today, scientific research overwhelmingly supports this concept:

The human body is composed of approximately 60–70% water.

Living cells rely on water to maintain their structure and facilitate chemical interactions.

Even theories about the origin of life—such as those involving early oceans or hydrothermal vents—place water at the center of life's emergence.

Astrobiology, a modern scientific field, searches for water as the primary indicator of potential life on other planets. Where water is found, scientists consider the possibility of life; where it is absent, life is considered highly unlikely.

Knowledge Ahead of Its Time

What makes this correspondence remarkable is not the complexity of the Qur'anic statement, but its accuracy. The verse does not claim how life evolved, nor does it describe biological mechanisms. It simply identifies water as the source of life—an assertion that science has only been able to confirm conclusively in recent history.

This precision avoids speculation while remaining consistent with future discovery.

Reflection and Meaning

The relationship between water and life illustrates a broader principle: fundamental truths often appear simple once understood, yet remain hidden until humanity develops the means to uncover them.

The Qur'an's reference to water as the origin of life serves as an invitation—not to abandon science, but to reflect on the harmony between natural discovery and ancient revelation.

Conclusion

Life begins, survives, and flourishes through water. This reality, now firmly established by modern science, was articulated in the Qur'an with remarkable clarity centuries before scientific confirmation was possible.

In recognizing water as the foundation of life, we are reminded that some truths transcend time—waiting patiently for humanity to discover their full meaning.

Chapter Seven

The Darkness in the Depths of the Oceans

For centuries, the depths of the oceans remained one of the greatest mysteries known to humankind. The human eye could not penetrate the vast darkness beneath the surface, and no technology existed to explore what lay beyond a few meters of water. Yet, more than fourteen centuries ago, the Qur'an described the deep sea with astonishing precision—long before submarines, sonar, or modern oceanography.

The Qur'an states:

"Or [their deeds are] like darknesses within a deep sea, covered by waves, upon which are waves, above which are clouds—darknesses, some of them upon others. When one puts out his hand, he can hardly see it."

(Qur'an 24:40)

This verse presents a layered and detailed image of the deep ocean: a deep sea, waves above waves, clouds, and multiple layers of darkness. Remarkably, each of these elements corresponds with discoveries made by modern marine science.

Multiple Layers of Darkness

Today, scientists know that sunlight does not penetrate the ocean evenly. As depth increases, light is absorbed gradually:

Red light disappears within the first few meters.

Orange and yellow lights vanish next.

Green and blue light fade as depth increases.

Eventually, complete darkness prevails in the deep ocean.

This results in multiple layers of darkness, exactly as described in the Qur'anic expression "darknesses, some of them upon others." Such knowledge was impossible to obtain through observation in the 7th century.

Waves Above Waves

Modern oceanography has also revealed the existence of internal waves—massive underwater waves that occur between layers of water with different densities. These internal waves exist beneath surface waves and are invisible to the human eye.

The Qur'an's phrase "waves, upon which are waves" aligns precisely with this phenomenon:

Surface waves driven by wind

Internal waves are moving deep beneath the surface

A Deep, Dark Sea

The verse refers specifically to a "deep sea" (bahr lujji), indicating waters of great depth. Scientists now classify deep oceans as regions where sunlight cannot reach, creating an environment of extreme darkness, pressure, and cold—conditions unknown to early civilizations.

A Knowledge Beyond Its Time

At the time the Qur'an was revealed, humans had no means to dive into deep oceans or study their physical properties. Describing layered darkness, internal waves, and the nature of the deep sea would have been beyond human capability.

This remarkable alignment between the Qur'an and modern science invites the reader to reflect deeply. It is not presented as a textbook of science, but as a sign—one that encourages humanity to seek truth, knowledge, and understanding of the universe.

THE DARKNESS IN THE DEPTHS OF THE OCEANS
"Or like darknesses in a deep sea, covered by waves, upon which are waves, above which are clouds—darknesses, some of them upon others."
Qur'an 24:40

Chapter Eight

The Sky as a Protected Ceiling

For most of human history, the sky was perceived as a vast, empty space above the Earth—beautiful, distant, and mysterious, yet largely passive. Only in recent centuries has science revealed that the sky is not merely empty space, but a complex and active system that plays a crucial role in sustaining life on our planet.

A Qur'anic Description

The Qur'an makes a concise yet profound statement:

"And We made the sky a protected ceiling, but they turn away from its signs."

(Qur'an 21:32)

This verse describes the sky not simply as a ceiling, but as one that is protected—a term that implies preservation, shielding, and safeguarding.

Modern Scientific Understanding

Today, modern science confirms that the Earth is surrounded by a carefully structured atmosphere that performs multiple protective functions essential for life:

Shielding from Meteoroids:

Millions of meteoroids enter Earth's atmosphere every day. Most of them burn up due to friction before reaching the surface, preventing catastrophic impacts.

Protection from Harmful Radiation:

The atmosphere, particularly the ozone layer, absorbs dangerous ultraviolet radiation from the Sun, protecting living organisms from severe biological damage.

Temperature Regulation:

The atmosphere helps maintain Earth's temperature within a narrow, life-supporting range by retaining heat and distributing it across the planet.

Retention of Essential Gases:

Gravity and atmospheric pressure prevent vital gases—such as oxygen and nitrogen—from escaping into space, allowing life to exist and evolve.

Without this protective system, Earth would resemble barren planets like Mars, incapable of sustaining complex life.

Knowledge Beyond Observation

At the time of the Qur'an's revelation, humans had no understanding of atmospheric layers, radiation, or space debris. The idea that the sky functions as a protective system would have been impossible to verify through observation alone.

Yet the Qur'anic description captures the essence of this reality with remarkable accuracy—without scientific jargon, and without speculation.

Precision Without Scientific Claims

It is important to note that the Qur'an does not describe how the atmosphere works in technical detail. Instead, it identifies its function. This functional description aligns perfectly with modern discoveries, while avoiding claims that could later be contradicted.

This balance between simplicity and precision is one of the most striking features of the Qur'anic approach to natural phenomena.

Reflection

The concept of the sky as a "protected ceiling" invites reflection rather than argument. It encourages humanity to look beyond appearances and recognize the unseen systems that make life possible.

What once appeared to be empty space is now known to be a finely balanced shield—silently protecting life on Earth, just as described centuries ago.

Conclusion

Modern science has revealed that the sky above us is not passive, but essential and protective. It guards the planet from harm, regulates its climate, and preserves the conditions necessary for life.

The Qur'an's description of the sky as a protected ceiling stands as another example of how ancient revelation aligns with truths that humanity would only come to understand much later—inviting thoughtful inquiry and reflection across generations.

THE SKY AS A
PROTECTED CEILING
“And We made the sky a intoitad ceilling...”
- Qur'an 21:32

Chapter Nine – The Breath of the Heaven

The Quran describes human experiences in ways that align with modern science, even in phenomena that were unknown 1400 years ago. One remarkable example is the feeling of chest tightness and difficulty in breathing at high altitudes, which we now understand is caused by a .decrease in oxygen concentration as one ascends in the atmosphere

So whoever Allah wants to guide – He expands his breast for Islam;“ and whoever He wants to misguide – He makes his breast tight and ”.constricted, as if he were ascending into the sky

(Al-An'am: 125)

Scientific Insight

Oxygen and Altitude: As we climb higher in the atmosphere, the air pressure decreases, reducing the amount of oxygen available for .breathing

Chest Constriction: This physiological effect makes the chest feel .tight and the breathing laborious, exactly as described in the verse

Connection with Guidance: The Quran uses this natural phenomenon as a metaphor: ease and openness in the chest symbolize .guidance, while tightness and discomfort symbolize misguidance

Why This is Remarkable

The Quran mentioned a phenomenon centuries before humans .understood the science of high altitudes and oxygen deficiency

The description is clear, precise, and relatable even to modern .readers

It blends physical reality with spiritual meaning, making it a .powerful lesson for science and faith

Chapter Ten – Wind: Nature's Pollinator

The Qur'an draws attention to the precise systems that govern the natural world, describing processes whose scientific explanations were only fully understood in modern times. One of these remarkable processes is the role of wind in the pollination of plants, a function essential to the continuation of life on Earth.

The Qur'an states:

"And We send the winds as pollinators, then We send down water from the sky, and We give life with it to a dead land. Thus is the resurrection."

(Qur'an 15:22)

Scientific Perspective

Wind Pollination:

Many plant species depend on wind to carry pollen from one plant to another, enabling fertilization and the production of seeds and fruits. This process, known today as anemophily, is fundamental to agriculture and natural ecosystems.

Sustaining Life:

Wind-driven pollination ensures the survival and reproduction of countless plant species, forming the foundation of food chains that support both humans and animals.

Connection Between Wind and Rain:

The verse also links wind with rainfall, highlighting a coordinated environmental system. Modern science confirms that winds play a critical role in cloud formation, precipitation, and the distribution of moisture across the Earth.

Why This Is Remarkable

The Qur'an described this biological and ecological function centuries before botany and environmental science explained it in detail.

The wording is direct and unambiguous, referring clearly to the fertilizing role of winds.

The verse presents nature as an integrated system, where wind, water, and life are interconnected in a precise and balanced order.

Reflection

Rather than presenting scientific theories, the Qur'an points to observable realities within nature and invites reflection upon them. The description of winds as pollinators reveals a level of insight that aligns strikingly with modern scientific knowledge.

This harmony between ancient revelation and contemporary discovery encourages thoughtful inquiry into the origins of such knowledge and the intricate design of the natural world.

Chapter Eleven – Iron: A Gift Sent Down to Earth

For most of human history, iron was known only as a powerful material—used in tools, weapons, and construction. Its true cosmic origin, however, remained completely unknown. Only in the modern scientific age did humanity discover a remarkable fact: iron did not originate on Earth.

A Clear Qur'anic Statement

The Qur'an states:

"And We sent down iron, in which there is great strength and benefits for humanity."

(Qur'an 57:25)

The expression "sent down" is direct and unambiguous. It does not describe iron as something merely created on Earth, but as something brought to it.

Modern Scientific Discovery

Astrophysics has revealed that iron is formed only under extreme conditions that do not exist on Earth. The temperatures and pressures required to create iron atoms can be found only inside massive stars.

When such stars reach the end of their life cycle, they explode in colossal events known as supernovae. During these explosions, heavy elements—especially iron—are forged and scattered across space.

Scientific evidence shows that:

Earth lacks the nuclear conditions necessary to produce iron.

Much of the iron found on Earth arrived via meteorites and cosmic debris.

Iron-rich asteroids collided with the early Earth, depositing vast quantities of this element.

These discoveries were confirmed only in the 20th century through advances in astronomy, nuclear physics, and space exploration.

Iron and Life on Earth

Iron is not only a cosmic element—it is essential for life:

It forms the core of hemoglobin, enabling oxygen transport in human blood.

It is critical for energy production in living cells.

Modern civilization depends on iron for infrastructure, technology, and industry.

Without iron, complex life and advanced civilization would be impossible.

Knowledge Beyond Human Reach

At the time of the Qur'an's revelation, humans had no knowledge of stellar explosions, nuclear fusion, or cosmic element formation. The idea that iron originated beyond Earth would have been unimaginable.

Yet the Qur'an described iron as something that was sent down, a statement that aligns precisely with modern scientific understanding.

Reflection

The Qur'an does not explain how iron was formed, nor does it describe supernovae or astrophysical processes. Instead, it presents a concise functional truth—one that science would only confirm many centuries later.

This convergence between revelation and discovery invites thoughtful reflection. It challenges the reader to consider how such knowledge could appear in a text revealed long before humanity could observe the stars in detail.

Conclusion

Iron is a cornerstone of both life and civilization. Modern science has shown that its origin lies not within Earth, but in the depths of the cosmos.

The Qur'an's description of iron as something "sent down" stands as a powerful example of alignment between ancient revelation and contemporary scientific discovery—offering yet another sign for those who seek truth with an open and rational mind.

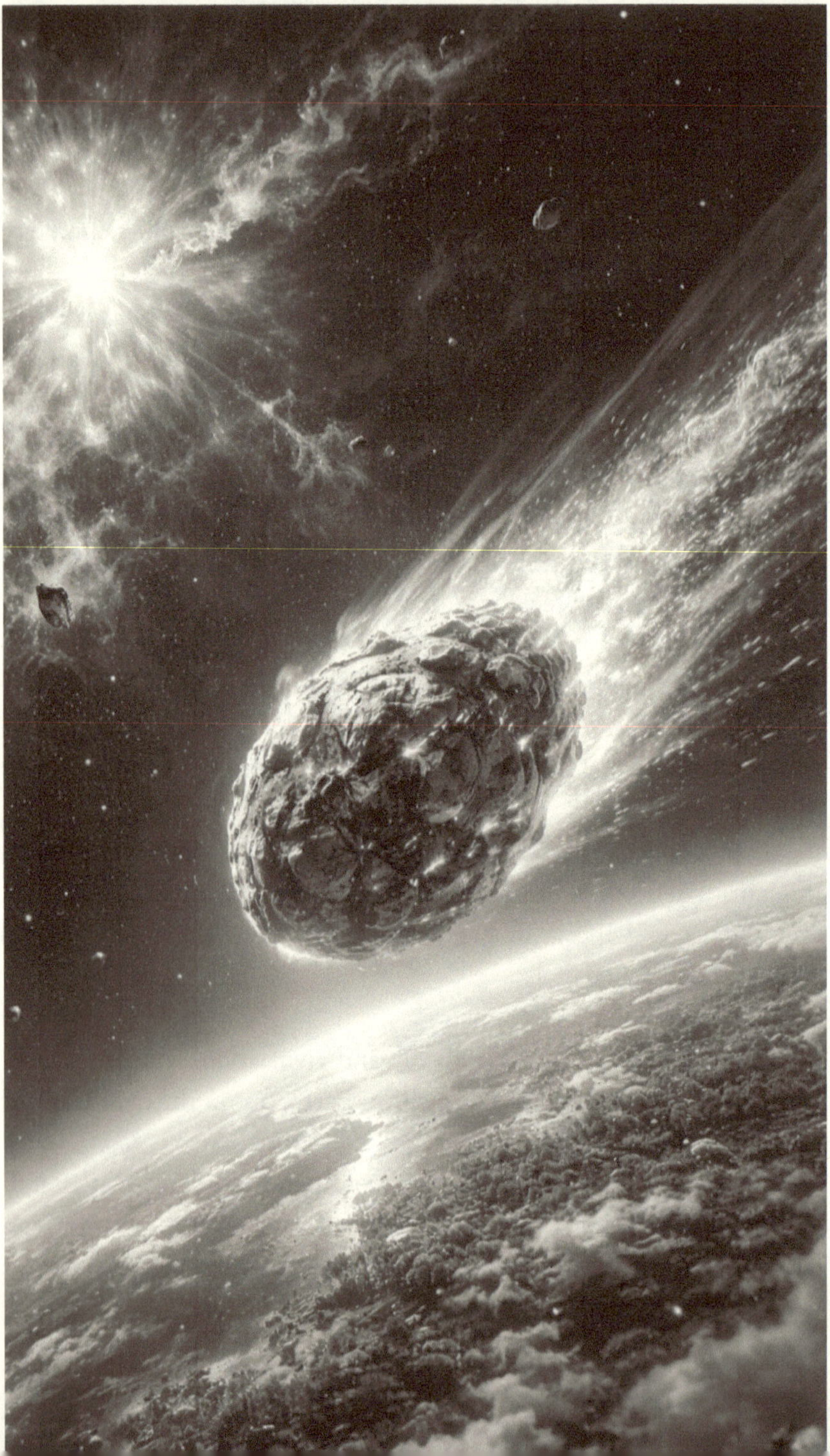

**Chapter Twelve
The Gates of the Sky**

For much of human history, the sky was perceived as an open and empty expanse. Ancient civilizations imagined it as a vast void above the Earth—limitless, unobstructed, and easily accessible in thought, even if not in practice. There was no concept of layers, barriers, or structural order. The heavens were "above."

This understanding remained dominant for centuries. Before the development of modern science, humanity had no knowledge of atmospheric pressure, radiation, gravitational limits, or the invisible forces that surround the planet. The sky appeared calm, open, and passive—an illusion created by human limitation.

Yet the Qur'an presents a strikingly different description.

"And even if We opened to them a gate from the sky, and they continued ascending through it..."

(Qur'an 15:14)

This verse introduces a powerful and precise idea: the sky has gates. It is not portrayed as an open emptiness, but as a domain with controlled access. Entry is not natural, random, or effortless. It requires a "gate."

At the time of this revelation, the concept of a structured sky would have been unimaginable. There were no aircraft, no rockets, no satellites, and no scientific tools to explore what lies above the clouds. Yet modern science has since revealed a reality that closely aligns with this description.

Today, we know that the Earth is surrounded by a complex and highly organized atmospheric system. The atmosphere is divided into distinct layers, each with unique properties:

The troposphere, where weather occurs

The stratosphere, containing the ozone layer

The mesosphere, where meteors burn

The thermosphere, with extreme temperatures and radiation

Beyond these layers lies outer space—a region that is not simply "above," but fundamentally hostile to life.

Far from being empty, the sky is dense with invisible constraints. Atmospheric pressure decreases rapidly with altitude. Oxygen becomes insufficient for human survival. Radiation intensifies. Temperatures fluctuate violently. Without advanced technology, the human body cannot endure ascent beyond a limited height.

Modern space travel has confirmed an essential truth: ascending into the sky is not possible without precise conditions. Rockets cannot rise at random angles. Launch windows are calculated to the second. Trajectories must align with Earth's rotation, gravity, and atmospheric resistance. Even a minor deviation can result in mission failure.

In scientific terms, engineers speak of windows, corridors, and paths—specific conditions under which passage is possible. Outside these conditions, ascent is blocked by physical laws.

In this context, the Qur'anic use of the word "gate" is remarkable. A gate implies:

Restriction

Control

Specific access points

Inability to enter freely

The verse does not say "if they ascended into the sky," but rather "if We opened to them a gate from the sky." Access is conditional. Permission is required. The sky is not open—it is guarded by laws.

The verse continues by suggesting that even if such a gate were opened, denial would persist. This highlights another dimension: the challenge is not merely physical, but intellectual. Evidence alone does not guarantee acceptance.

From a modern scientific perspective, ascent into the sky is an extraordinary achievement. It demands knowledge, technology, protection, and exact precision. For most of human history, such ascent was impossible—not by choice, but by design.

The convergence between this ancient description and modern discovery is difficult to ignore. Without scientific terminology, the Qur'an presents a conceptual framework that reflects reality: the sky is a structured system, not a void.

This chapter does not claim that the Qur'an is a science textbook. Rather, it invites reflection on how a text revealed in a pre-scientific age speaks with such accuracy about the natural order—using language that remains meaningful even centuries later.

The sky, as we now know, is a constructed realm. It has limits. It has layers. And it has gates.

**Chapter Thirteen

A Universe Without Cracks**

For much of human history, the universe was perceived as vast but poorly understood. It appeared mysterious, distant, and possibly chaotic. Stars moved, planets revolved, and the sky stretched endlessly above—but no one truly understood whether this immense structure was stable or fragile, ordered or accidental.

Only in recent centuries has modern science revealed a profound and unexpected truth:

The universe is exact, coherent, and structurally consistent.

The Qur'an draws attention to this reality strikingly and directly:

"[He is the One] who created seven heavens in layers. You do not see in the creation of the Most Merciful any inconsistency. So return your vision—do you see any cracks?"

(Qur'an 67:3)

This verse does not describe the universe poetically or emotionally. Instead, it presents a rational challenge:

observe carefully—do you find any flaw, inconsistency, or structural failure?

From Chaos to Precision

Before the rise of modern physics, many assumed that the universe was either eternal chaos or a loosely arranged system. There was no concept of finely balanced forces or universal constants. The idea that the cosmos operates under a unified, precise framework was entirely absent.

Today, scientific discovery tells a very different story.

A Finely Tuned Cosmos

Modern cosmology has demonstrated that the universe is governed by a set of fundamental constants—such as gravity, the speed of light, and the strength of nuclear forces—that are calibrated with astonishing precision.

If any of these values were altered even slightly:

Stars would not form

Galaxies would collapse or disperse

Life would be impossible

Physicists refer to this phenomenon as fine-tuning. The universe, despite its vast scale and continuous expansion, remains stable and coherent.

No Cracks in the Fabric of Reality

The Qur'anic term translated as "cracks" refers not merely to physical fractures, but to structural flaws or inconsistencies. From a modern scientific perspective, this aligns closely with what researchers observe:

No contradictions in the laws of physics

No breakdown in the fabric of spacetime

No instability that threatens the universe as a whole

Even as galaxies move apart and stars are born and destroyed, the underlying structure remains intact.

Expansion Without Collapse

One of the most remarkable discoveries of modern science is that the universe is expanding. Intuitively, such expansion might suggest eventual tearing or instability. Yet this does not occur.

The universe expands without losing coherence. Its laws remain constant everywhere. This balance—between motion and stability—is one of the greatest puzzles in cosmology.

The Qur'an's invitation to "return the vision" reflects a timeless observation:

the more closely one examines the universe, the more order appears—not less.

Knowledge Beyond an Ancient World

At the time of the Qur'an's revelation, humanity lacked telescopes, equations, or any understanding of cosmic structure. There was no framework through which such a conclusion could be reached by observation or speculation alone.

Yet the text directs attention to a reality that science would only confirm centuries later:

the universe is not fractured, random, or flawed—it is structurally sound.

Reflection

This chapter does not claim that the Qur'an teaches physics. Rather, it highlights how a pre-scientific text draws attention to a truth that aligns with the deepest discoveries of modern science.

The universe, when examined carefully, does not reveal cracks in its design—but consistency, balance, and order.

The question then is not merely scientific, but philosophical:

How did such insight appear in a text revealed long before humanity could observe the cosmos in detail?

Chapter Fourteen – The Skin: The Center of Sensation

The human body is a remarkable creation, and one of its most sophisticated organs is the skin. The Qur'an describes its role in a striking verse:

"Every time they are touched, we replace their skins with new ones, so that they may taste the punishment."

(Qur'an 4:56)

At first glance, the verse may seem symbolic. Yet modern science has revealed a profound truth: the skin is not merely a protective covering, but the primary sensory organ of the body.

1. The Skin as a Sensory Organ

The skin

contains millions of nerve endings.

These receptors detect:

Touch

Pressure

Temperature

Pain

It allows humans to feel the environment, respond to danger, and experience pleasure or discomfort.

2. Layers and Renewal

Human skin consists of multiple layers: epidermis, dermis, and subcutaneous tissue.

The epidermis is constantly renewing itself—dead cells are shed, and new cells form.

This continuous regeneration is echoed in the Qur'an's description: "We replace their skins with new ones."

3. The Science of Pain and Sensation

Pain is processed through the skin's nerve endings.

Experiments show that the skin can detect a single touch or a tiny temperature change, sending signals instantly to the brain.

In terms of the Qur'an, the verse highlights how sensation is central to the experience of reward or punishment, which modern neuroscience confirms: without skin, the perception of physical reality would vanish.

4. Reflection

The Qur'an pointed to the functional importance of the skin centuries before modern anatomy.

What we now know about pain, touch, and the complex structure of the skin aligns with the description in the verse.

This chapter invites readers to reflect on how the human body is precisely designed and how even seemingly simple features, like the skin, are essential for life and experience.

5. Conclusion

The human skin is much more than a covering. It is a gateway to sensation, a record of touch and feeling, and a mirror of the body's interaction with the world. The Qur'an's reference to skin replacement highlights both its sensitivity and its remarkable capacity for renewal, a truth only confirmed by science centuries later.

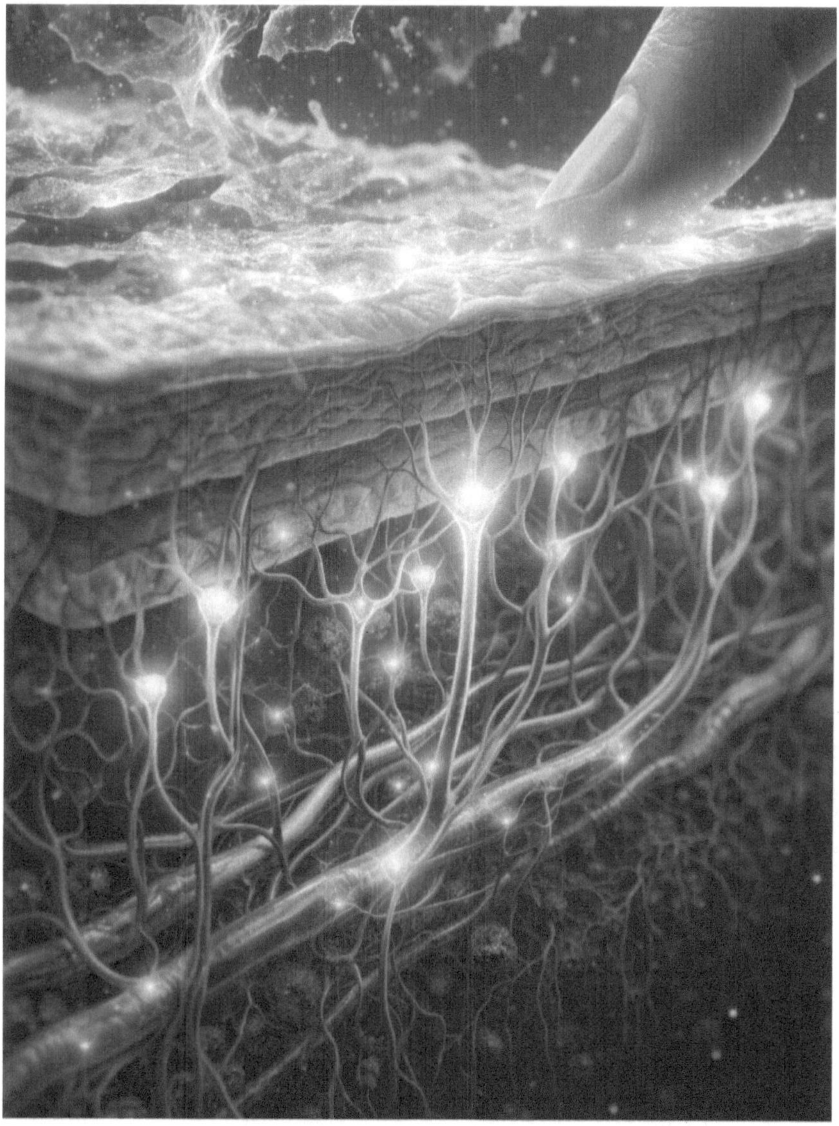

Chapter Fifteen

Pure Milk from Between Waste and Blood

A Precise Biological Process Described in the Qur'an

Qur'anic Statement

The Qur'an states:

"And indeed, in livestock there is a lesson for you.

We give you to drink from what is in their bellies—

from between waste and blood—pure milk, pleasant to those who drink it."

(Qur'an, 16:66)

This verse presents a remarkably precise description of milk production, using clear and concrete terms: waste, blood, and pure milk. The wording is neither poetic nor symbolic. Instead, it describes a biological process that remained completely unknown to humanity until the development of modern physiology.

What Was Known Before Modern Science

For most of human history, the internal workings of animal bodies were a mystery. People observed that animals ate plants and produced milk, but how this transformation occurred was entirely unknown. There were no microscopes, no understanding of digestion, no knowledge of blood circulation, and no concept of specialized glands.

Even ancient physicians believed milk was simply transformed blood or a substance stored in the stomach. These explanations were speculative and incorrect.

Yet the Qur'an describes milk as emerging between digested food waste and blood—an astonishingly accurate statement.

Modern Scientific Understanding of Milk Production

Today, biology explains milk production through several well-documented stages:

Digestion (Waste – "Farth")

Food consumed by an animal is digested in the stomach and intestines. The remaining material becomes waste, while nutrients are absorbed through the intestinal walls.

Absorption into the Blood

Nutrients such as amino acids, fatty acids, sugars, vitamins, and minerals pass into the bloodstream. At this stage, the blood acts as a transport system, carrying these substances throughout the body.

Milk Formation in the Mammary Glands

The mammary glands extract specific components from the blood and re-synthesize them into milk through a highly regulated biochemical process.

Importantly:

Milk is not blood

Milk is not digested waste

Milk is a newly formed substance

Separation and Purification

Harmful substances, toxins, and waste products are filtered out. The result is a sterile, nutritious liquid—milk—perfectly suited for consumption.

"From Between Waste and Blood" — A Striking Accuracy

The Qur'an does not claim that milk comes from waste or from blood.

It states that milk comes from between them.

This is exactly what modern science confirms:

Milk originates after digestion

Its raw materials travel through the blood

Its final production occurs in an independent system (mammary glands)

This precise distinction could not have been known in the 7th century.

"Pure Milk, Pleasant to the Drinkers"

The verse also highlights two additional properties:

Pure:

Milk is naturally sterile when produced. It contains antibacterial agents and immune-supporting compounds, especially in its natural state.

Pleasant and palatable:

Despite originating from processes involving waste and blood, milk is clean, nourishing, and agreeable—something that science recognizes as a remarkable biological achievement.

Why This Cannot Be Explained by Guesswork

Several points make coincidence or speculation extremely unlikely:

No ancient culture described milk production accurately

No visible observation could reveal this process

The wording aligns precisely with modern physiology

The verse avoids common historical misconceptions

This is not a vague statement that can be reinterpreted later. It is specific, restrained, and correct.

A Question for the Truth-Seeker

If this knowledge was unknown to physicians, philosophers, and scientists for centuries, an honest question arises:

How could an unlettered man in the Arabian desert describe this process with such precision?

The Qur'an presents this not as a scientific textbook, but as a sign—an invitation to reflect, question, and search for truth.

Conclusion

The production of milk is one of the most sophisticated biological processes in nature. The Qur'an's concise description aligns perfectly with discoveries made more than a thousand years later.

This verse stands as a clear example of scientific accuracy in the Qur'an—

not exaggerated, not forced, but plainly stated.

For the sincere reader, it is not merely information,

but an open door to deeper reflection.

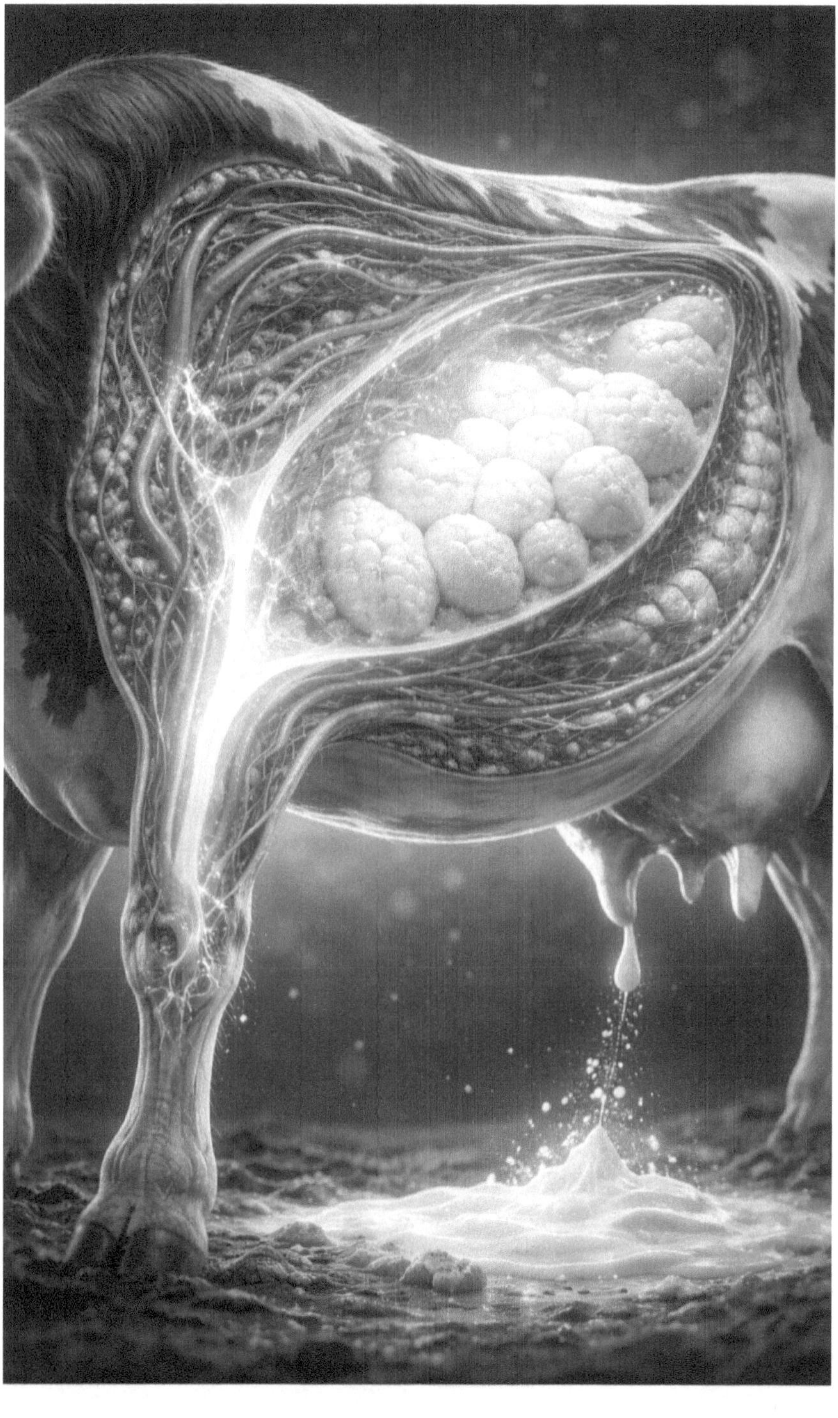

Chapter Sixteen

Honey: Healing for Humanity

A Natural Remedy Described in the Qur'an

Qur'anic Statement

The Qur'an states:

"From their bellies comes a drink of varying colors, in which there is healing for people."

(Qur'an, 16:69)

This verse speaks about a specific natural substance—honey—and makes a clear, measured statement: "in it is healing for people."

The wording is precise. It does not claim that honey cures every disease, nor does it exaggerate its effects. Instead, it presents honey as a source of healing, a description that aligns remarkably well with modern scientific findings.

What Was Known Before Modern Science

For thousands of years, honey has been used in traditional medicine across many civilizations. It was applied to wounds, consumed for illness, and valued as a nourishing substance. However, the reasons behind its healing properties were unknown. Its effectiveness was observed, but never scientifically explained.

Without knowledge of microbiology, enzymes, or chemical compounds, ancient use of honey relied entirely on experience rather than understanding.

Modern Scientific Discoveries

Today, science has uncovered why honey possesses healing properties:

Antibacterial Action

Honey contains natural antibacterial compounds, including hydrogen peroxide and, in some varieties, methylglyoxal. These substances inhibit the growth of harmful bacteria.

Wound Healing

Medical-grade honey is used in hospitals to treat burns and chronic wounds. It promotes tissue regeneration, reduces infection, and accelerates healing.

Anti-Inflammatory and Antioxidant Properties

Honey helps reduce inflammation and contains antioxidants that protect cells from damage.

Effectiveness Against Resistant Bacteria

Research has shown that honey can be effective against bacteria that are resistant to conventional antibiotics, making it valuable in modern medical treatment.

These properties are now well-documented through laboratory research and clinical studies.

"Healing for People" — A Precise Description

The Qur'an does not describe honey as a miracle cure. Instead, it states that within it exists healing. This careful wording is scientifically accurate:

Honey supports healing processes

It benefits certain conditions

Its effectiveness depends on usage, quality, and context

Such precision avoids false claims while remaining consistent with verified medical knowledge.

A Substance of Remarkable Origin

Honey is produced by bees through a complex and highly organized process:

Nectar is collected from flowers

Enzymes are added during digestion

Moisture is reduced through evaporation

The final product becomes a stable, long-lasting substance with medicinal value

This transformation results in a natural remedy unlike any other—produced without human intervention, yet optimized for preservation and healing.

Why This Knowledge Is Significant

At the time of the Qur'an's revelation, there was no scientific explanation for why honey healed wounds or fought infection. The statement that honey contains healing properties could not have been derived from chemical analysis or medical theory.

Yet modern science confirms that this description was accurate.

Reflection

This verse invites thoughtful reflection rather than blind acceptance. It points to a natural phenomenon and allows discovery to unfold over time. As science advances, the depth of meaning within the text becomes increasingly clear.

Honey stands as a bridge between ancient observation and modern scientific understanding—an everyday substance that reveals extraordinary properties when studied carefully.

Conclusion

Honey is more than a sweet substance. It is a biologically active material with proven healing effects. The Qur'an's measured description of honey as containing healing aligns precisely with modern medical knowledge.

For the sincere reader, this convergence offers more than information—it offers an invitation to reflect on the origin of such insight and the harmony between natural law and revelation.

Chapter Seventeen

The Camel: A Marvel of Creation

Reflecting on the Qur'an

The Qur'an invites reflection:

"Do they not look at the camels, how they are created?"

(Al-Ghashiyah: 17)

This verse directs the reader to observe a creature that is both visible and tangible, yet its design contains complex and astonishing adaptations that were unknown to humanity for centuries.

Unique Features of the Camel

Modern science confirms that camels are among the most well-adapted animals on Earth. Their design is perfectly suited to extreme desert environments:

Sustaining Life in Arid Conditions

Camels can survive weeks without water.

They conserve water efficiently and can drink up to 40 liters at once when water is available.

Fat Storage and Energy Management

The hump stores fat, not water, which provides energy over long journeys.

Fat metabolism produces water internally—a natural system for desert survival.

Temperature and Heat Regulation

Camels tolerate extreme heat during the day and cold at night.

Their body temperature fluctuates to minimize sweating and water loss.

Specialized Circulatory System

Camels' red blood cells are oval, enabling them to flow smoothly even when blood becomes concentrated due to dehydration.

Efficient Digestive System

Able to digest coarse, thorny desert plants.

Water absorption and nutrient extraction are highly optimized.

Why This Observation Matters

The Qur'an does not merely say: "look at animals"—it asks specifically to reflect on the camel's creation. This is remarkable because:

Ancient people had no knowledge of internal anatomy or physiology.

Yet the statement draws attention to observable and measurable adaptations.

Today, every aspect of the camel's survival has been studied and confirmed by science, showing the precision of the Qur'anic statement.

Invitation to Reflection

For the reader seeking truth, the verse presents:

A visible miracle: an animal that can endure extreme conditions.

An opportunity to understand complex biological systems without instruments or microscopes.

Evidence of design and wisdom in creation, inviting contemplation rather than blind belief.

Conclusion

The camel is more than a desert beast. It is a masterpiece of biological engineering, optimized for survival in one of the harshest environments on Earth.

The Qur'an's call to observe and reflect on the camel is both a scientific and spiritual invitation, encouraging the reader to connect observation with reason, and to appreciate the depth of natural wisdom.

Final Conclusion

Conclusion

In this book, the purpose has never been to subject the Qur'an to the changing standards of modern science, nor to measure its truth by human discoveries that remain limited, provisional, and constantly evolving. For those who believe in it, the Qur'an is not a human text awaiting verification, nor an idea open to correction or revision.

It is the Word of the Creator, Exalted and Glorious, and as such, it stands as a standard in itself.

Therefore, the correct question is not: Does the Qur'an agree with modern science?

Rather, the more honest question is: How has modern science, after centuries of observation and experimentation, come to discover realities that the Qur'an pointed to from the very beginning?

What has been presented in these pages is not an attempt to "prove" the Qur'an, for the Qur'an does not require proof. Instead, it is a presentation of remarkable points of convergence between contemporary scientific discoveries and a text revealed more than fourteen centuries ago—at a time when no microscopes, laboratories, telescopes, or scientific instruments existed.

As Muslims, we hold no doubt whatsoever regarding what the Qur'an has conveyed. Our faith is not built upon scientific conclusions, nor does it rise or fall with them. Our certainty precedes science and transcends it.

Yet we also recognize that the Qur'an addresses the human intellect, invites reflection, and repeatedly calls upon people to observe the universe, the human being, and the laws that govern existence.

For this reason, this book is directed primarily toward those who do not believe in the Qur'an, and toward anyone sincerely searching for truth. It is not a call to blind belief, nor an attempt at intellectual coercion. Rather, it is a calm and reasoned invitation to reconsider—to pause and reflect with fairness and openness.

How could a text that emerged in a simple environment, in an era devoid of scientific knowledge, contain insights that align so closely with discoveries made only in modern times?

These pages do not demand uncritical acceptance. They offer an opportunity to think before it is too late—to grant this divine text a fair hearing, free from prejudice and preconception, with an open mind and an honest heart.

In the end, the Qur'an remains a witness to itself, and truth reveals itself to those who genuinely seek it.

"We will show them Our signs in the horizons and within themselves until it becomes clear to them that it is the Truth."

(Qur'an 41:53)

Don't miss out!

Visit the website below and you can sign up to receive emails whenever Messaoudi Abou nouh publishes a new book. There's no charge and no obligation.

https://books2read.com/r/B-A-RZBFF-UFIAJ

www.ingramcontent.com/pod-product-compliance
Lightning Source LLC
LaVergne TN
LVHW090429210926
848480LV00002B/755